U0013761

Life Lessons
I Learned
from My Dog

Illustrated by Emma Block

汪想
賴在你身邊

沒有人是一座孤島，
汪星人的100則陪伴哲學

艾瑪‧布洛克 —— 繪   陳采瑛 —— 譯

 抱緊推薦

　　養狗，不只是照顧一條生命，你會發現牠們給自己的還要多更多！

　　「Q 將」慵懶的態度吸引了不少粉絲，那股淡定萌的力量每天也都療癒你我的心。「PU 將」的加入讓我們學習著如何平衡相處，每一個生命，都有牠它存在的意義與價值。

　　《汪想賴在你身邊》作者的每張繪圖，畫的不僅是毛小孩每天的日常，更畫出了牠們對主人的愛與包容。我們每天牽著牠們散步，牠們同時也牽著我們的心，彼此相依陪伴。

　　　　　　　　　　—— Q 娘「柴犬 Q 將」臉書粉絲團主

陪伴，是最珍貴的禮物。

一開始擁有狗的時候，我以為彼此的關係就只是一方付出、另一方接受的上對下關係，但隨著時間堆疊，卻漸漸發現牠們帶來的禮物遠比想像中還要多更多。人寵關係從來都不是單方面的給予，而是雙向的交流。狗狗用牠們的一生來教導我們如何成為更好的人，牠們教會我們如何與另一個生命相處和負責。生命的歷程中總有高低起伏，但只有在低落與不順遂的時候才能知道誰對你不離不棄，而狗狗永遠是我們最忠實的夥伴。

書中的一百則汪星人語錄，道出了狗狗們對我們的愛、包容及陪伴，真摯推薦給大家一起來細細品嘗。

——聖文 & 米菇「跟著有其甜」臉書粉絲團主

狗真的很讚。

這本書裡面充滿簡單但溫暖的插圖，有養狗的人看過一定都能會心一笑。有時人類很難做到或看見的事情，在狗身上卻很容易就發現了，我們藉由養狗來給我們力量，並在這之中學習到最簡單快樂的幸福。

每個人有能力都應該養條狗，光看著狗睡覺，就覺得時間好像變慢了，非常地愜意。

——寶總監 知名動物圖文作家

這本書獻給我的父親，保羅‧布洛克，
還有我們家的愛犬，山咪（2003-2019）

在我的成長過程裡一直都有狗狗陪伴。我爸媽養的第一隻狗，麥斯，只比我大幾個月。那時我們兩個都在長牙齒，經常一起在客廳地板上咬書。之後我爸媽養了另外一隻狗，山咪，十五年來都是我們家的愛犬。

成為大人後，每次回父母家就表示一定有隻毛茸茸的朋友在門廳等我，星期天下午我們花很長的時間在海邊散步，晚上一起窩在沙發上，到處沾滿了狗毛。我一直都愛狗，也會永遠愛狗。牠們是天生的樂觀主義者，呆萌、愛玩、寬容，以及永遠忠誠，這就是我們可以從牠們身上學到些什麼的原因。

艾瑪・布洛克

永遠保持熱忱

Always be enthusiastic

用愛克服恐懼

Overcome fear with love

不記仇

Don't hold grudges

每天都要玩

Play every day

接納
你自己

Accept yourself

表現關懷

Show compassion

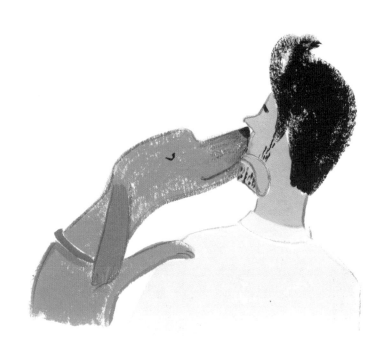

告訴你愛的人你有多在乎他們

Show your loved ones how much you care

享受旅行

Enjoy the journey

無條件去愛

Love unconditionally

開心時
就跳起來

Jump for joy when you're happy

問候你所愛的人

Greet the ones you love

你絕對不會老到玩不動

You are never too old to play

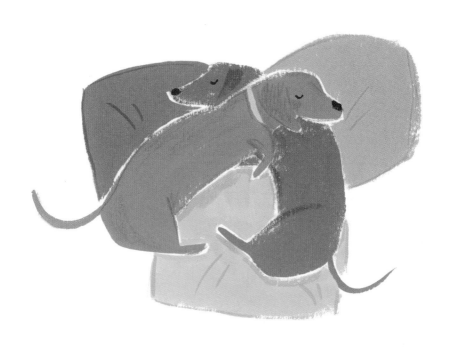

欣賞彼此

Appreciate one another

大人有大量

Forgive easily

用你的步調過生活

Take life at your own pace

品嘗你的食物

Savour your food

要忠誠

Be loyal

享受靜默

Enjoy the silence

活在當下

Live in the moment

保持
專注

Stay focused

交朋友可以很簡單

Friendship can be easy

喝很多水

Drink lots of water

做會讓你感到快樂的事

Do the things that make you happy

各種工作都值得做好

Any job is worth doing well

保持活力

Stay active

善待自己

Treat yourself

追求你覺得重要的事物

Go after the things that are important to you

總是要
目光
交會

Always make eye contact

揮別錯誤向前走

Move past your mistakes

花很多時間在戶外

Spend lots of time outside

不要怕弄髒

Don't be afraid to get messy

沉浸在寧靜的片刻

Enjoy the quiet moments

欣賞
風景

Enjoy the view

相信你最初的感覺

Trust your instincts

挖出埋藏的寶物

Dig for buried treasure

絕對不要偽裝成
不像自己的樣子
Never pretend to be something you're not

睡到飽

Get enough sleep

外表
可以
唬人

Looks can be deceiving

讓 怒 氣 消 散

Let your anger go

每天早上都要伸展

Stretch every morning

想要有空間的時候就吠叫

Growl when you want some space

生活中簡單的事物最棒

The simple things in life are the best

停下來
傾聽

Stop and listen

愛，不求回報

Love without expectation

劃好你的地盤

Establish your boundaries

每天都要散步

Go for walks every day

保持好奇心

Stay curious

嘗試各種事物至少一次

Try anything once

保持良好關係

Stay connected

不要只想到自己

It's not just about you

別擔心
旁人
怎麼看你

Don't worry what others think of you

交付信任，也要有智慧

Place your trust wisely

難受的時間不會持續到永遠

Tough times never last

活出
你的本色

Rock your own style

給每個人……

Give everybody...

一個機會

...a chance

不開心的時候不用隱藏

Be upfront when you're not happy

樂於接受讚美

Accept compliments well

每天
都試著
去分享

Try to share every day

適應新環境

Adapt to new surroundings

愛才是最重要的

Love is all that matters

保持良好的……

Keep a good...

生活平衡
...life balance

要正向積極

Be positive

雖然覺得害怕但還是拚了

Feel the fear and do it anyway

不說長道短

Don't gossip

生活可以更單純……

Life is simpler...

當你只追求你想要的

...when you ask for what you want

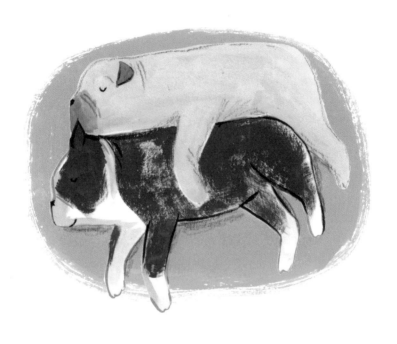

沒有人
是一座
孤島

No one is an island

人生苦短，沒時間傷悲

Life is too short to be sad

保有童心

Don't grow up

好朋友，一輩子

A best friend is for life

眼神接觸最重要

Eye contact is everything

跟著你的福氣走

Follow your bliss

小狗
也可以有
大夢想

Being small doesn't mean you can't dream big

信心滿滿地為自己加油

Carry yourself with confidence

不從封面評判一本書

Don't judge a book by its cover

心能裝下多少才是關鍵

It's the size of your heart that matters

「變化」為生命添加風味

Variety is the spice of life

放輕鬆，
別顧慮那麼多

Don't take yourself too seriously

朋友形形色色

Friends come in all shapes and sizes

先行動，再道歉

Act first, apologize later

準備周到

Be prepared

珍惜
耍廢的日子

Cherish lazy days

樂於探險……

Embrace adventure...

但接近未知事物時
要保持警戒

...but approach the unknown with caution

別因為他人的批評而沮喪

Don't let the critics get you down

請求協助沒什麼好丟臉

There's no shame in asking for help

一見鍾情是真的

Love at first sight is real

心指向它想要的

The heart wants what it wants

完美……
來自於……練習！

Practice...makes...perfect!

抓住所有機會

Seize all opportunities

緊盯著你的目標

Keep focused on your goal

有時候
傾聽就是
最好的陪伴

Sometimes just listening is best

我們相像的地方
其實比不像的還多

We are more alike than different

只要多練習，就能走到哪睡到哪

With practice, you can nap anywhere

感到滿足快樂的關鍵
就在你自己
You yourself hold the key to contentment

有愛，怎麼看都美

Beauty is in the eye of the beholder

有時候

融入周遭

也很好

Sometimes it's okay to blend in

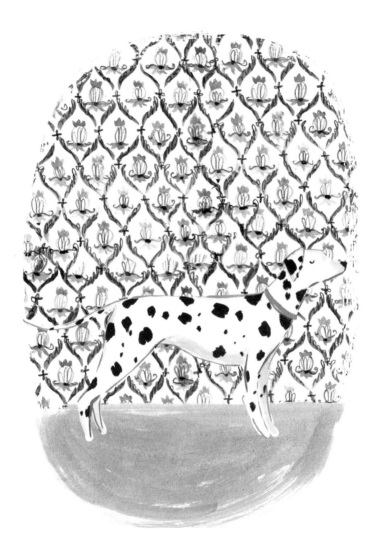

堅持會有回報

Persistence pays off

| 繪者簡介 |

艾瑪・布洛克（Emma Block）

英國插畫藝術家，擅長美術編輯、裝幀與品牌設計。作品風格深受大眾喜愛，曾與英國哈洛德百貨、時尚設計師奧蘭・凱利等聯名合作，亦開設水彩繪圖與藝術字工作坊，出版水彩教學專書。繪畫的靈感都來自於日常生活、老照片、復古衣飾、黃金年代的爵士樂，還有她的臘腸狗。

| 譯者簡介 |

陳采瑛

畢業於中央大學英文所，偶爾客串當譯者。譯過許多以動物為主角的圖文書，包含《走路要有喵態度》、《一顆海龜蛋的神奇旅程》、《燕子的旅行》、《頑固的鱷魚奶奶》、《歡迎光臨蟲蟲旅館》、《鳥巢大追蹤》等。

汪想賴在你身邊

沒有人是一座孤島，汪星人的100則陪伴哲學

圖——艾瑪·布洛克（Emma Block）
文——Michael O'Mara Books Ltd
譯——陳采瑛

責任編輯——陳嬿守
主　　編——林孜勲
封面設計——謝佳穎
內頁排版——陳春惠
行銷企劃——鍾曼靈
出版一部總編輯暨總監——王明雪

發行人——王榮文
出版發行——遠流出版事業股份有限公司　100臺北市南昌路二段81號6樓
　　　　　　電話／(02)2392-6899　傳真／(02)2392-6658　郵撥／0189456-1
著作權顧問——蕭雄淋律師
□2020年4月1日 初版一刷

定價——新臺幣320元（缺頁或破損的書，請寄回更換）
有著作權·侵害必究　Printed in Taiwan
ISBN——978-957-32-8741-4

YL-■遠流博識網 http://www.ylib.com　E-mail: ylib@ylib.com
遠流粉絲團 https://www.facebook.com/ylibfans

國家圖書館出版品預行編目 (CIP) 資料

汪想賴在你身邊：沒有人是一座孤島，汪星人的100則陪伴哲學
　/ 艾瑪・布洛克（Emma Block）著 ; 陳采瑛譯. -- 初版. -- 臺北
　市 : 遠流, 2020.04
　　　面 ;　　公分
　　譯自 : Life Lessons I Learned from my Dog
　　ISBN 978-957-32-8741-4(精裝)

1.犬　2.通俗作品

437.35　　　　　　　　　　　　　　　　109002609